BEYOND THE LIMITATIONS OF
LIVINC ENDLESS YESTERDAYS

await infinite possibilities today.

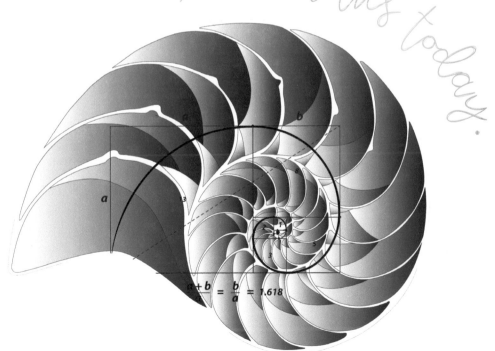

$$\frac{a+b}{a} = \frac{b}{a} = 1.618$$

First published in 2020
by MMH Press
Waikiki, WA 6169

Interior and cover design: Cassandra Neece

National Library of Australia Catalogue-in-Publication data:
Quantum Thinking/Adrea L. Peters

ISBN: (hc) 978-0-6489376-1-6
ISBN: (e)   978-0-6489376-2-3

QUANTUM THINKING

Factoring in Possibility

ADREA L. PETERS

In gratitude, and in grace,

FROM ME TO YOU.

TABLE OF CONTENTS

Welcome

I forgot I wrote this book. I had scribbled these sayings, now offerings, into a crumpled spiral. The notebook is a purple Top Flight I bought at a grocery store years before. The edges are frayed, worn and soft. The metal spine still strong and prepared to protect these treasures for years to come. This particular spiral also contains rough sketches I drew for my house lay out, the paint colors I selected for it, budgets, oh so many budgets, goals, dreams, ideas for Truitt Book 3, and a long love letter I wrote to my father who died in April of 2017, the year that notebook went everywhere with me.

I wrote these down to capture the essence of Truitt as I started to map Book 3. For those who don't know my work, Truitt Skye is a character who lives in a world of quantum physics that I call The City on the Sea. As I began to capture them, purely to keep myself straight for my readers, there was a bit of a shift, from a fictional bible to a whispered from above potency. They seemed somehow. . . massive, and utterly obvious. So I set them aside and moved onto the next thing.

On July 23, 2020, I was having a chat with Rebecca Gibson. She's a bit of a powerhouse and heaven-sent if you ask me. A true catter for Truitt fans out there. She said, "What's this quote book thing? Golden ticket?" Frowning and clueless, I was like, "Ahhhh... dunno what you're talking about." Profound, right? Yep. That's me. We poked around at it a bit (she's not one to let you slide) and finally I guessed, "Maybe it's *Higher Mathematics*? It's these little blurbs about life. Quantum physics."
She grinned wide, as she does when she's pleased, and said, "That's it. Publish it immediately. Put Quantum in the title. And get it done before the end of the year."

So here we are.

I originally dubbed it, *Higher Mathematics: Factoring in Humanity,* because I thought it would be a companion book to the Truitt Skye series. That title felt perfect coming from Truitt. Now that it's coming from me, we've got to reverse it to: *Quantum Thinkinφ: Factoring in Possibility.* Nothing lights me up more than your possibility. It is endless. I hope you can feel that. It is everything.

A central component to Life, which Truitt represents, is that we are all the same, that the many come from the one, that all life is connected. No matter how cut up or destroyed things are, we cannot break apart from all that is. We are all children of Nature.

The math of Nature is φ, pronounced fee. It represents divine proportion, which is the foundation for a three-dimensional world (our world) as expressed in the Fibonacci sequence, which is often referred to as the Golden Mean, Golden Ratio or Golden Sequence. This sequence gives math (the language of unification) to just about everything from leaves to flowers to faces to bodies to temples and tombs to gas and ether and roots and music and starfish and seashells. Even to Consciousness, otherwise known as, Thinking, can be recognized with the Golden Ratio. The One or Whole or Collective (Divinity or The Universe or Humanity) to the Many (Humans).

My intention with *Quantum Thinkinφ,* is for it to feel like you are holding The Golden Ticket. That you and the Universe are one for a perfect moment as you read the page you opened to. I love to imagine this little book sitting on your desk or countertop or nightstand waiting for you to receive what you need instantly. It travels with you, tucked in your satchel for a quick peek during an annoying meeting. My dream? Right there by the loo for a quick peruse whilst you take care of things. I do aspire.

Please know it's written with every part of me loving every part of you across all time-space continuums. May it bring you more and more and more with every read.

All my love,

adrea

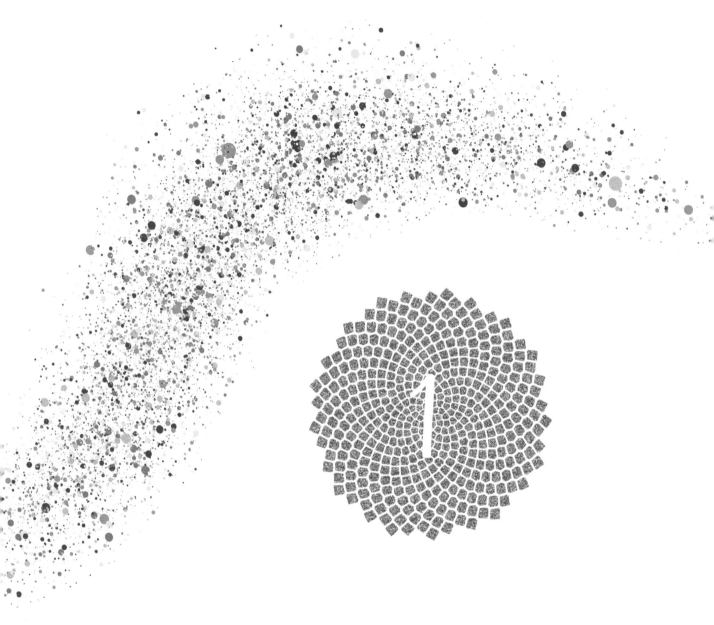

Permission

Permission to be human was and always will be granted. You are never doing anything wrong. Every*thing* and every*one* adds value. Always.

Oneness

We all share the same life story. Many births, many deaths. Not one. It is a mistake to think you're only born once, and only die once. Not true biologically. Much less, emotionally or experientially.

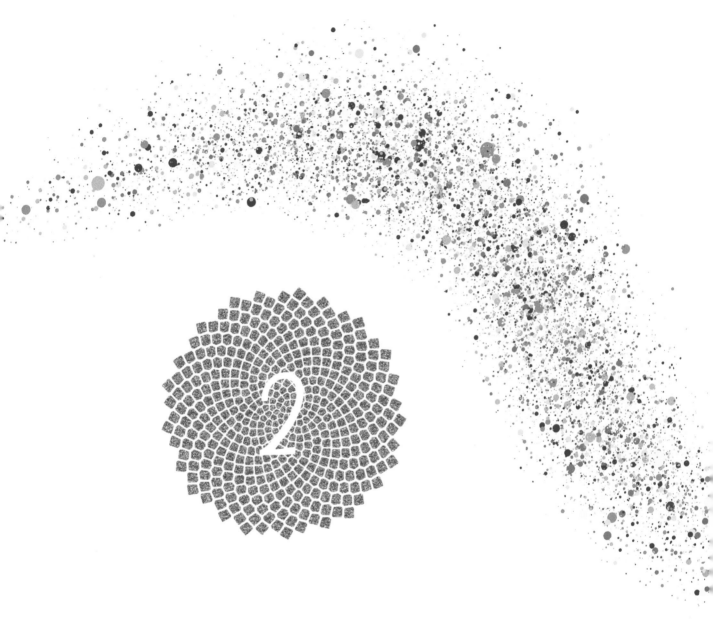

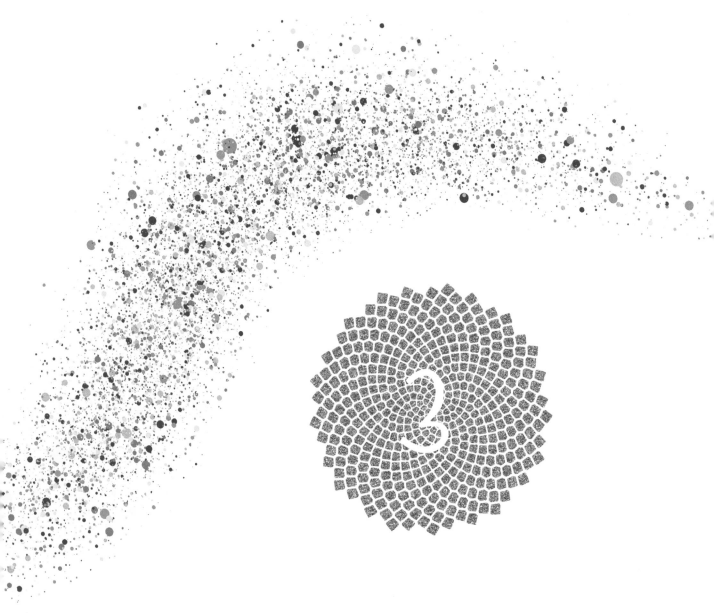

Equanimity

Story is the formula of human life. It levels the playing field. Set your narrative. Or others will do it for you.

Superpowers

Listening. Hearing. Thinking. Feeling. Laughing. Loving.
Being human is all you need to be a Superhero.

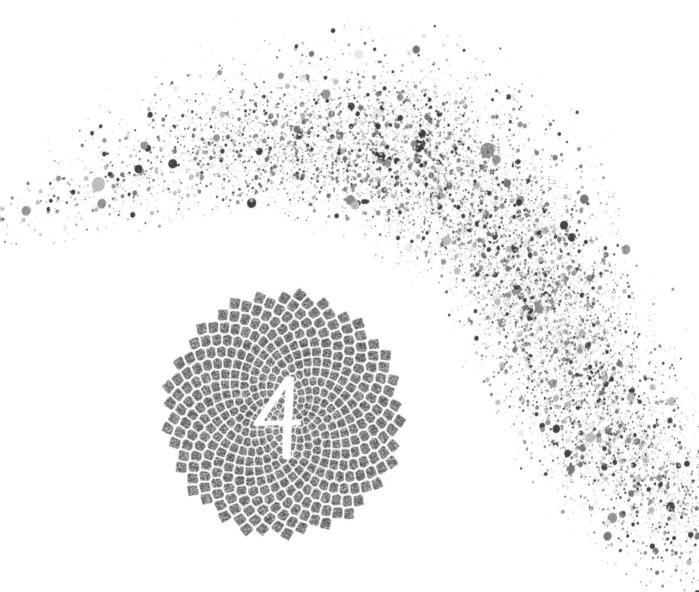

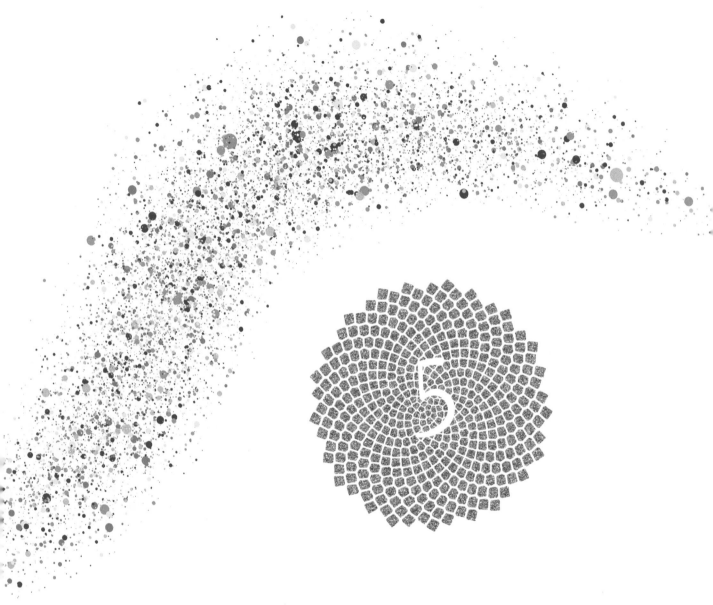

Theme

Themes are energetic waves breathing life into us. Each unique theme holds your deepest desire, and will elude you until you own each and every one.

Heard

We *want* to be seen and known, loved and honored. We **need** to be heard. And the only one who can **actually**, *without question,* HEAR you, is you.

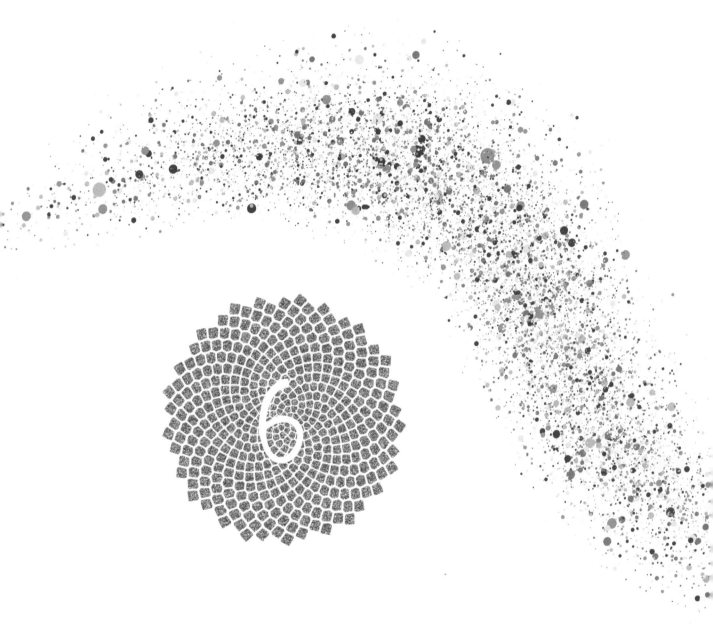

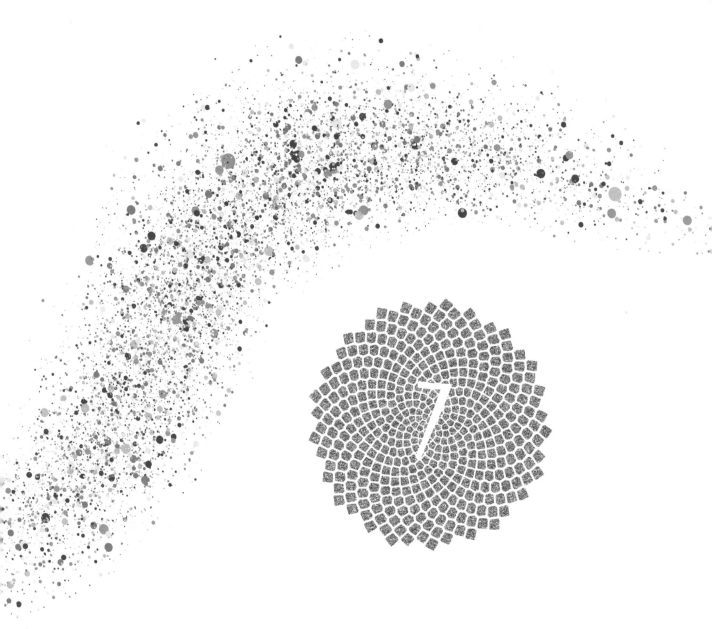

Unconditional

The only kind of love is unconditional love. No bounds, no rules, does not require certain behavior or ways of being, and has no limitations whatsoever. This love can never, ever be felt from another. You must, if you do nothing else with your entire life, love yourself unconditionally before you can receive or offer your love to another. The math simply does not work without $a=you$.

Alone

Your life is a journey of one. We intersect with millions, billions of other beings, but you are it. We must blaze our trail alone, amidst the millions. This does not mean others do not matter, or that they are better or worse than you, or that you are meant to be alone, quite the opposite. But you must first find your way on your own. And you must wish the same for others, so they feel the support and love of someone's desire for them to thrive on their own. They are you and you are they. Knowing you is knowing they. No exclusion, only inclusion.

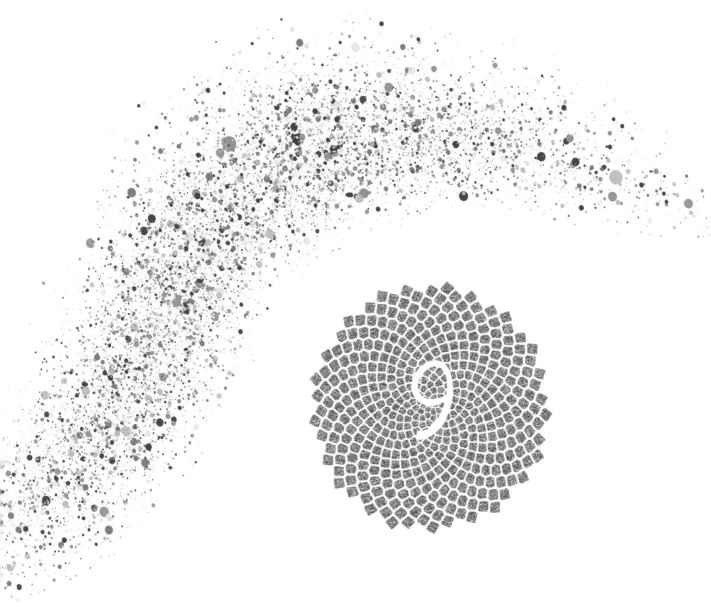

Personal

Everything is Personal. You are a PERSON! Of course it's personal. Take it personally, you beautiful person you! Feel it. Gently learn how to express yourself. Screw up. Offend. Hurt yourself and others. Let us not be afraid to make mistakes. Let us be more afraid of *not* respecting and expressing who we are: a person! Life revolves around us being a person with other persons. Only when we take it personally, do we to see that negative, harsh, rash behavior is incongruent to our need to hear and love ourselves unconditionally.

Kindness

When we start with Kindness as our base, as our constant, we have given ourselves a foundation of such strength, we will never need to act in any other way.

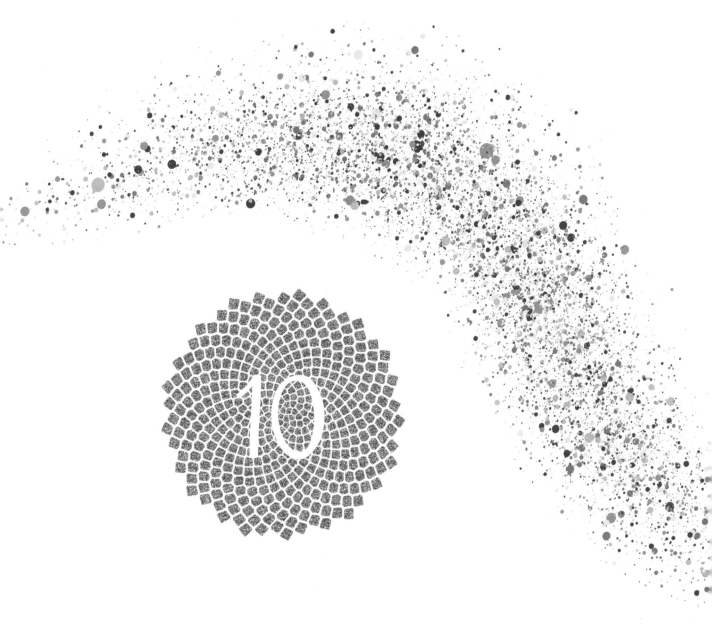

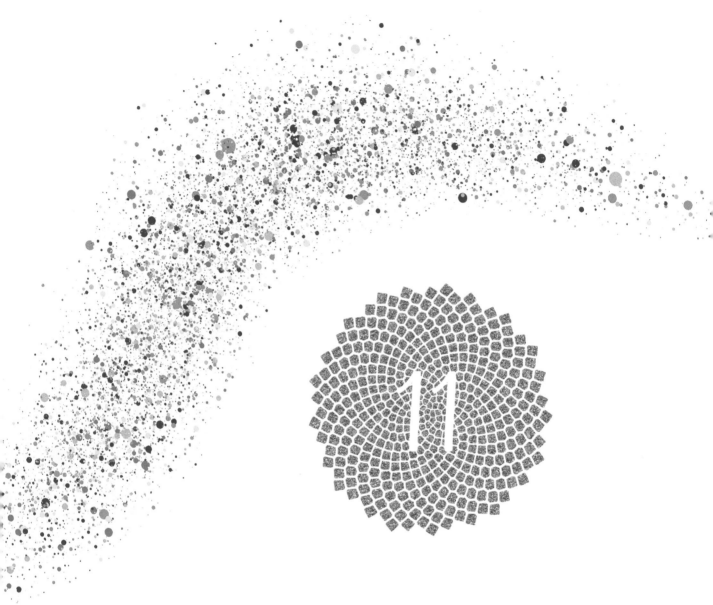

Surrender

We have become afraid of lulls, gaps, quiet, silence, slowness, delays and obstacles in our way. If there is too much space, we panic. Our threshold for quiet is shockingly low, sometimes seconds. We need to see "results" now. When we feel like things aren't progressing "fast" enough, we feel we must do something—anything. This includes sabotaging ourselves by taking a job we don't want, partnering with someone we do not love, saying something we do not mean or eating or drinking something we do not need. We've become allergic to letting things in our life sit there untouched and unattended. We no longer think things through. We react. And the consequence is stunting our ability to evolve, especially emotionally, arguably the most compelling ingredient to being human. When we surrender to the quiet, the lulls, the gaps, the silence, slowness, delays, and obstacles, things arrive right on time.

Praise

Praise has become cliché. Therefore, praise is lost. We cannot know what we have done well, without know what we have not done well. It is good to compare so that preference and excellence can emerge. Praise must be given with thought and with purpose. As should criticism.

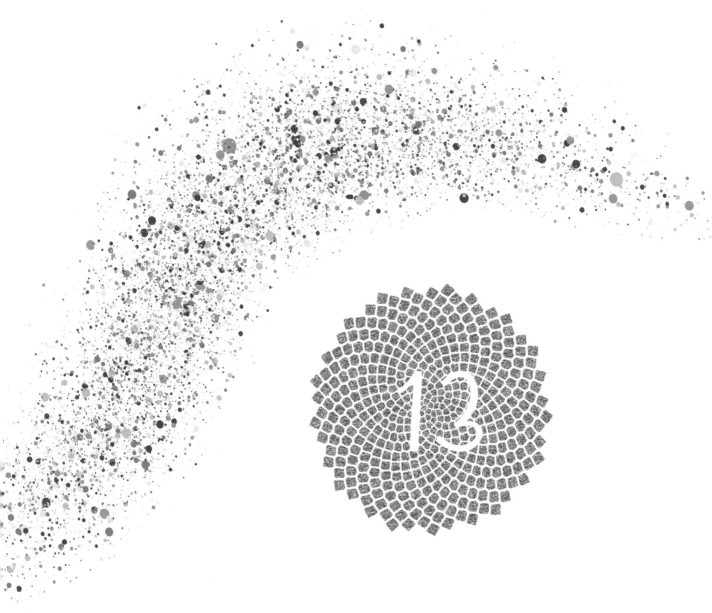

Vibration

Vibration is the manager of our body, of our well-being, of how we perceive the world and how the world perceives us. Vibration is not a woo woo thing. Nor it is not a spiritual thing. It is real. It is what and who you actually are. It speaks far louder than words and is always ten steps ahead of us.

Believing

The biggest human flaw we are currently, and collectively, suffering from is believing that the global state of affairs has nothing to do with our own *individual* believing. We are mistaken in believing that we are not part of the whole, of all humanity. We are. I don't need to tell you are human. Or a member of humanity. Yet, you don't believe me when I say that you absolutely and positively impact humanity by what you think and feel and believe. Moment to moment. Nanosecond to nanosecond. Your thoughts (energy) convert into matter. They matter. They have serious power. What we think about is what we get. Not just in our own lives. In everyone's lives. Nothing is off limits. We can create, or destroy, everything. And it has *nothing* to do with taking action. It has everything to do with believing.

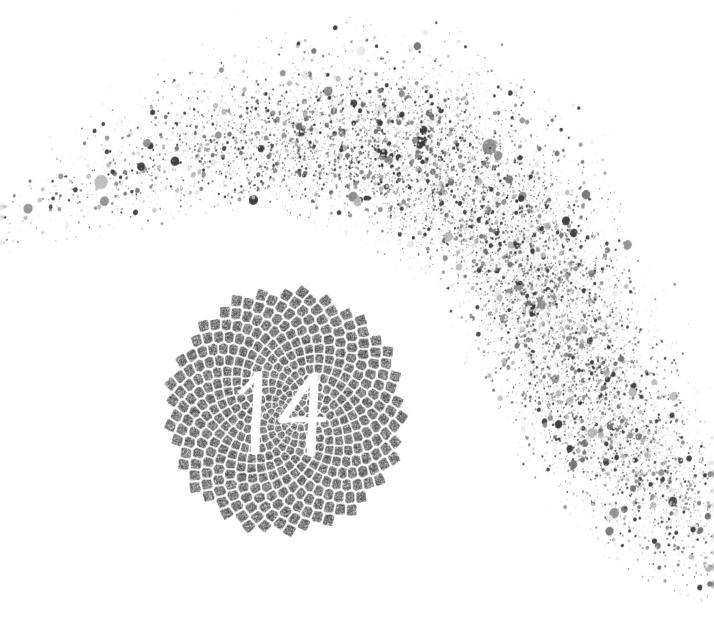

14

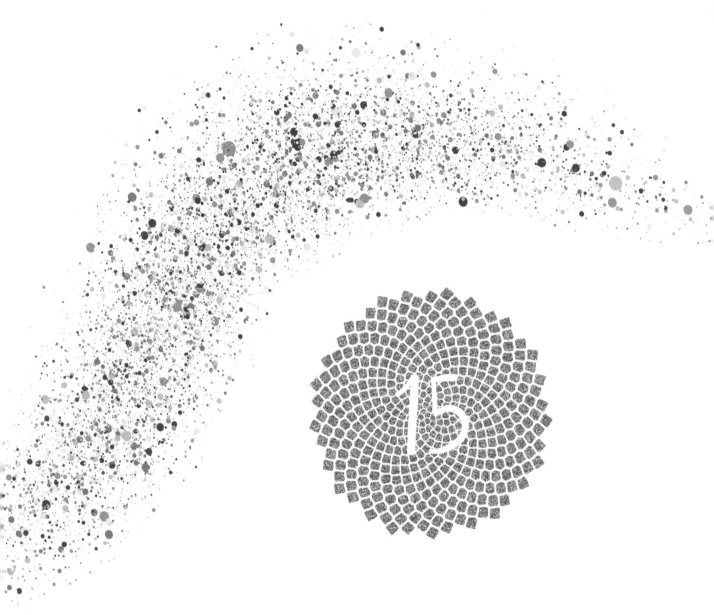

Perspective

Perspective is the fountain of youth. We don't have to hold onto the way things are. We don't have to accept our reality. The only thing real from our past is our perspective of it. *What If* are the magic words you've been looking for. What if you looked at this or that in a different way? The only thing we can change about anything is our perspective on it. Oh the power of perspective. Hone it. Mold it. Your perspective is your reality.

Commodities

Commodities are no longer gold, silver, pork bellies, corn or wheat. They are Sleep, Responsibility, Honesty, Interest, Time, Peace and Self-Awareness.

16

Focus

Energy moves in one direction: to or toward where attention is given, otherwise known as focus. Cold water becomes hot with energy/focus of fire/electricity. Good needs Bad. Love requires Hate. Bravery cannot be without Fear. Happiness is bound to Despair. Everything exists on the same plane, all inextricably connected. You can shorten the traverse from the negative to the positive with focus.

Individualism

We have shifted from growing as Individuals, to conforming to patterns and archetypes. You are unique. You are a one of a kind. Yet we run from the journey of Becoming who we are. Instead we want to be like them. Why not try being you?

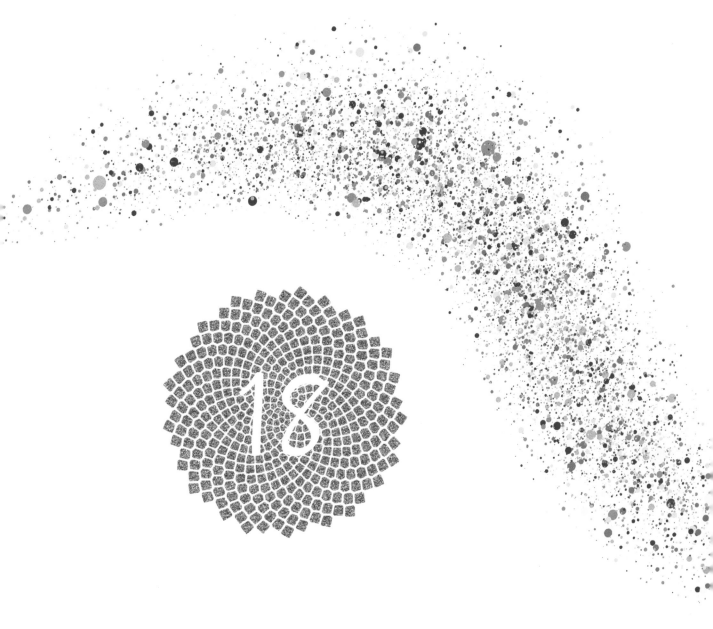

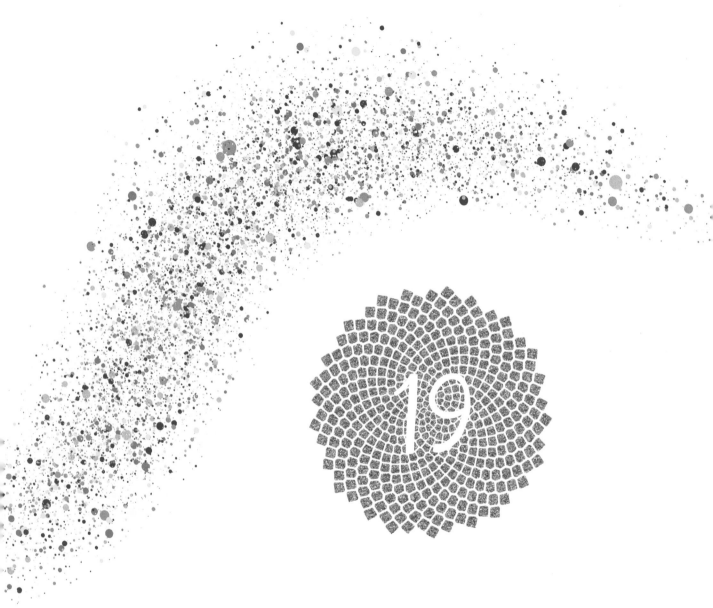

Coincidence

Coincidence is not Karma. Coincidence is connection. It is a reflection of your thinking, of your feeling, of your wanting, of your questions, your interests, your perspective. It is proof that you get what you think about.

Death

Death is not The End. We are only ever becoming more. There are no endings. Energy cannot be created or destroyed. Thank you, Rudolf Clausius! Energy only transforms or changes. It does not end. You do not end.

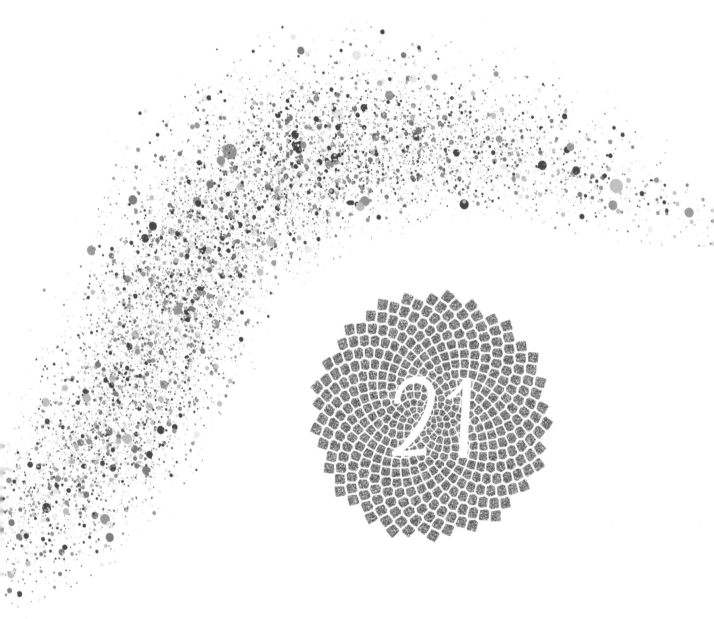

Difference

Comparison is only valuable when we are celebrating our differences. Celebrating similarities is a constant and not noteworthy. In fact, it is stifling. If we are only in search of sameness, we cease to grow. Evolution relies on mutation, anomalies and asymmetries to perpetually inspire growth and adaptation.

Truth

Truth is not collective. It is an individual choice to determine what is true and what is false, for you and only you. We can rely on our physical responses, resonance within our cells, bones, and blood to know what is true. Truth has become, and may have always been, an opinion, a belief. A truth may be an irrefutable fact, but that is up to you. What is true for you, may not be true for others. This is excellent. Diversity, difference, and disagreement are fundamental to growth and evolution. Homogenous, one-way thinking and living, lead to inertia, stagnation and extinction.

Loss

If you are lucky, you will lose EVERYTHING and eventually (hopefully quickly) realize you lost nothing at all. It is impossible to truly appreciate without devastation. Of course I don't wish this upon anyone, but if it does happen, you will see the gift of it—as long as the loss doesn't consume you. When we lose it all, we gain a new degree of wholeness from within, a closeness to ourselves that fortifies us.

Generosity

Give it all away! Give what you know, who you are, your genius, your skills, your love. And ask for nothing in return. Make room for the abundance. Create space for more of what you desire. We need to stop separating work or earning from who we are. Compartmentalizing ourselves means we live fragmented lives, which serves no one and leads to lies and hurts and emotional dread and pain. You cannot be anything but you. The math doesn't work. You can only be you. Share you. From there, everything is possible.

War

War, in all forms, is unnecessary. Violence is never needed. Anger, revenge, worry, fear are all figments of the imagining mind. These are the things raping the Earth. She will outlast us all, no matter how many wars we create. Once we stop the nonsense, we would see that clearly.

feeling

Feel before you think. You actually do this already, but the transition of emotion to thought is so fast, it typically goes unnoticed. Feel your way to everything. The Golden Question: How do I feel now? Constantly checking in with ourselves and tending to our emotional answers could change the world in unimaginably short order. Don't think. Don't tell. Shh. Feel. Your. Way. There. You know what should happen based on the pit in your stomach, the elation in your voice, the dread giving weight to your body, the unprovoked smile, the yawn, the pain, the nodding you didn't intend. Let your feelings lead the way.

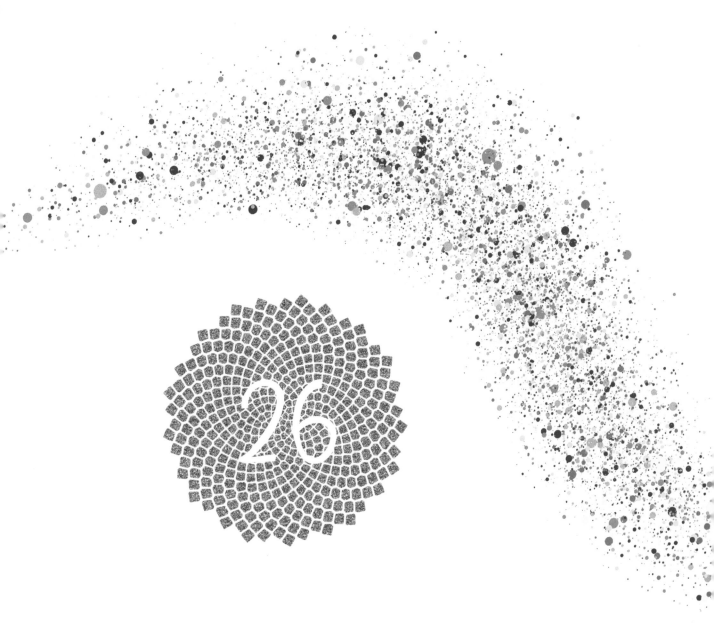

Thoughts

Your mind is not to be trusted unless, and until, your thoughts are trustworthy.

Mirrors

You don't have to worry that you don't know how you feel. A mirror will arrive and reveal it. When you are contented, others will smile at you, say hello, help you in a "surprising" way. Angry, upset, agitated? Fights will race to you, fender benders, rudeness, mishaps are all on their way. Everything that ever happens is a reflection of you.

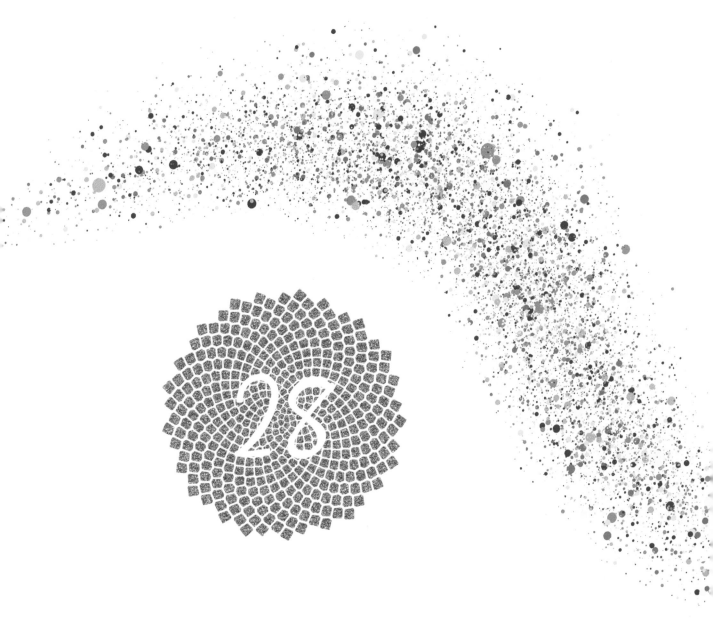

Just

Just is a 4-letter-word unless used to describe time or justice. Please stop using this word immediately. The emotion behind it is justification. You needn't justify anything to anyone ever. Diminishing yourself, diminishes everything.

Break Through

The best habit you can form is breaking through anything and everything that you're letting stop you from living your life to the fullest.

Responsibility

Give up Victimhood. Take full responsibility for everything in your life. Love it. Cherish it. Take responsibility for the good. As well as the bad. All of it. You are the owner of your life. No matter what has happened, no matter what you have endured, survived, overcome and dealt with, own it because no one can set you free but you. You may as well give up the struggle. You are the only one holding you back.

Love

Love is our anchor. I don't believe we are capable of "un-loving" someone. When we love, we love. Period. Love can shift based on experience. As we discover more about ourselves, love adjusts. Hate, Fear, Angst are transient. Love is not. Fear, hate, and sadness, always fall backwards to love. Emotional gravity is as natural as physical gravity.

Interest

Interest is the fuel of the soul. It is a gateway to connection. What interest you, you follow, and follow, and follow. Curiosity is not quite enough. It's the scent that piques your interest. You have to be *interested* to act. Many of us spend a great portion in our life doing things we aren't even curious about, much less interested in. This is why we feel tired, disconnected and separate. Find interest, find connection! Find life! Such a difference a little interest can make.

Let Go

You cannot convince anyone of anything. Cease all efforts immediately.

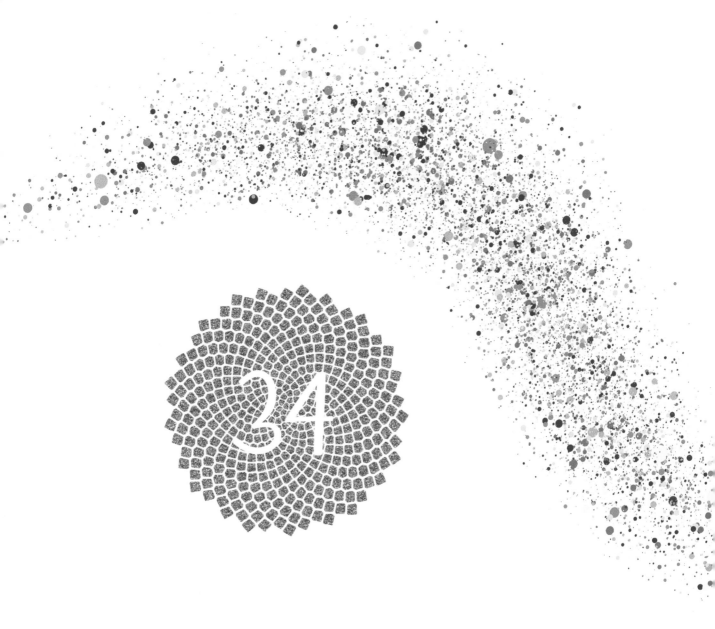

Plenty

You have all you need. Always have, always will. There is nothing you can do to be fulfilled by anyone but you. They cannot fill you with confidence, only you can. They cannot fill you with love, only you can. They cannot fill you with sadness, only you can. You are the cup. You fill you.

Time

Time is malleable. It is grounded in relativity. Think you need time? Add perspective and watch time bend to your will.

Relationships

We cannot know who we are without the constant feedback of others, otherwise known as mirrors. We become because of them. They are everything to us, and we are everything to them. Everyone needs everyone.

Intimacy

Intimacy is what we truly crave, yet it only takes a fraction of deceit to crack intimacy, making it as fragile as it is formidable. It is a connection so pure, so exquisite and real that we are nearly enveloped by the depth of love we achieve, yet completely free to experience every breath of it. Intimacy is oneness between two people. It is the Golden Mean. Divine Proportion. Everything all at once and then some more.

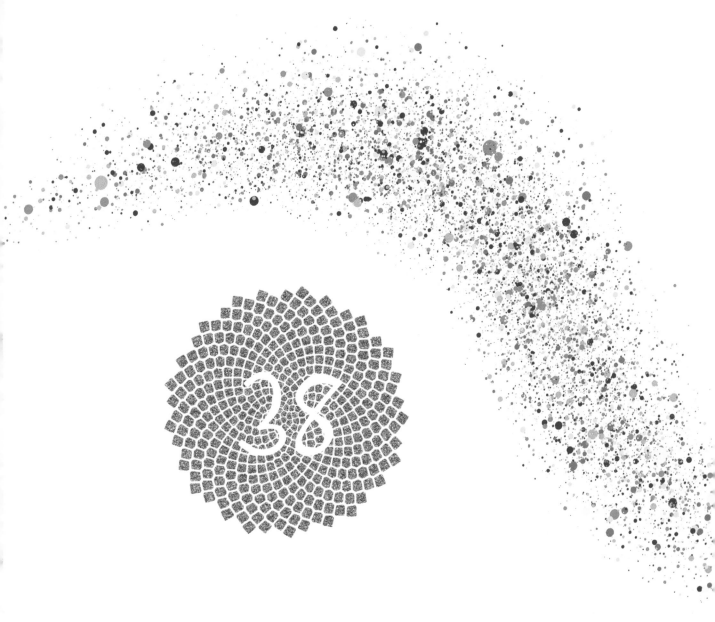

ACKNOWLEDGEMENTS

My dear readers, you are the reason for all that I create and
I would not be me without all of you. Thank you.

Cass. Your talent and support bring my ideas and creations to life.
You amaze me daily. Thank you.

Karen. You know. I know you know you mean the absolute world
to me because you help me soar. Thank you.

Rebecca. You keep me on my toes reaching for more. You are
exquisite. Thank you.

Miche. Remember when this was a wild WT idea? Holy shizzle it
was real. I love you, babe! Thank you.

Anj. Beautiful you. Thank you, sweet soul.

Mom. Dad. Ann. Scott. Tish. Sophie. Isabel. I love you. Thank you.

ABOUT THE AUTHOR

Adrea is an author-shareholder with MMH Press. She trained as a journalist, novelist and screenwriter. She earned a Bachelor of Science from the University of Colorado at Boulder and Master of Arts in Fiction Writing from Seton Hill University.

She lives amongst the trees in Vermont with her boys, Skye and Fig.

Please visit her website at *www.adreapeters.com* for more information.

OTHER TITLES BY ADREA L. PETERS

CPSIA information can be obtained
at www.ICGtesting.com
Printed in the USA
BVHW021954070920
588250BV00010B/56